"I loved the straight-forward and easy to follow presentation. I witnessed how it delivers results."
Peter Geraets

*"Statistical Replenishment as presented is a **breakthrough** in the control of the raw material and finished product flow,"* REM

"The positive impact of the stability improvement generated by Statistical Replenishment will be experienced by every organization along the supply chain" GLM

Ron Mueller

Stress Free™ Supply Chain Solutions
Flowing production and high productivity

By: **Ron Mueller**

Around the World Publishing LLC
4914 Cooper Road Suite 144
Cincinnati, Ohio 45242-9998

ISBN 13: 978-1-68223-121-0

Distributed by Ingram
Cover Picture By: Tassel78|Dreamstime.com
Cover Design By: Ron Mueller and Gordon Miller

Ron Mueller

SUPPLY CHAIN CONTENTS

DEDICATION
To all those who are willing to try some
breakthrough thinking and apply
supply chain flow.

ACKNOWLEDGMENTS

To all the people contributing to
keeping materials and finished products flowing
out to the customer

TECHNICAL EDITOR:

Gordon Miller P. E.

Introduction

The concepts in Stress Free™ Supply Chain Solutions are a combination of breakthrough, proven concept extrapolation and proven concept efficient and coordinated application.

The role of the leader must change to be that of supply chain orchestration. The Supply Chain leader must envision a supply chain organization that rewards the seamless flow of materials, information, and money to the beat of the purchasing customer.

They must envision the participants to a loss and waste free way to deliver the work required to get the desired product into the hands of the consumer.

These supply chain leaders must develop their capability to be hands on root cause problem solving coaches. They must be willing to learn each and every required type of work along the supply chain and then they must hold the organization to the standards required to maintain and improve the supply chain.

They must do something different with the supply chain if they expect a breakthrough. They must recognize that a breakthrough in performance is possible and it is an attractive and desirable business goal.

Stress FreeTM Supply Chain Solutions is the guide to a supply chain organization that achieves a 3-5% margin point improvement, a 40% reduction in inventory and greater than 20% improvement in productivity, an OEE equal to or greater than 85%, increased throughput, a high performance supply chain organization and customer service satisfaction of 99.97%.

A halo effect is the significant reduction

in both quality defects, and serious safety incidents.

Statistical Product Replenishment, Material Flow and Production Stability and Supply Chain High-Performance Organizational Design are among the key new and different concepts presented in this book that will yield dramatic supply chain performance improvement.

Statistical Product Replenishment

It is the key concept in freeing up trapped cash.

The Toyota PULL system is not feasible for most supply chains. Statistical Product Replenishment however provides a unique and immediately applicable way for product production to closely meet the customer demand.

The supply chain is the backbone of a business. It must remain fully functional during the transformation. The proposed transformation must yield substantial breakthrough gains. The change benefits must overwhelm the significant effort that change implies.

The concept of Statistical Product Replenishment is a breakthrough concept that has been proven to yield thirty to sixty percent inventory reduction in specific products being managed by leading production planning programs. A production control advisor tool provides guidance that is then utilized to manage production.

The application approach provided in this book provides a risk-free way to test the concept and once the approach is validated the transition to flow can be rapidly achieved.

Flow: The rate that the raw material flows from the supplier to the final customer. All systems have flow. Most flows are stop and go. The approach presented in this book guides the reader to a continuous flow system

The constant flow quantity, *flow pitch,* the standard daily production schedule and statistically determined response bands reduce the constant change in the production scheduling process.

Flow Pitch: Think of this as the size of the wave that is moving along the supply chain. The limiter to the size of the wave and period of the wave is what is defined and *flow pitch.*

An increase in operational stability is experienced within weeks of the application of statistically controlled production. The entire supply chain experiences a lessening of variability. Within three months an entirely different stable flowing supply chain begins to emerge.

Trapped cash is reduced enough to pay for all the improvement work along the supply chain. The financials allow for new mutually beneficial organizational relationships to be established among all the supply chain organizations and put into rapid practice.

Material Flow and Production Stability

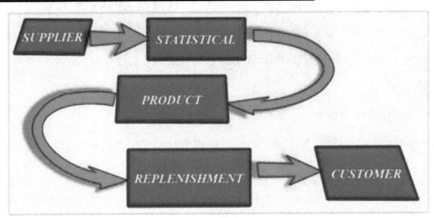

Material Flow

Material flow will follow the behavior of the statistically controlled product replenishment. The timing and quantity of material movement will match the statistically controlled production process.

This coordination will require new relationships with key raw material suppliers. These suppliers will experience a significant reduction in their trapped cash. Renegotiating the contracts with suppliers is normally a very positive experience.

Production Stability

Production stability provides the basis for growing the skills of all the people along the supply chain, creating stable, reliable equipment.

Daily Management, a high-performance teamwork organization and root cause problem solving capability form the basis for the development of people that bring the production equipment to an OEE (**O**verall **E**quipment **E**ffectiveness) that is greater than 85%.

A daily management process focused on fixing problems to root cause provides a bedrock foundation. This is a requirement for longer term continuous improvement. The breakthrough here is the ability to increase the problem-solving skill and capability of the operating team members.

Stress Free™ Manufacturing Solutions and its accompanying excel workbook enables line teams to solve problems at a graduate engineer's level.

Supply Chain High Performance Organization

The Supply Chain High-Performance Organization design builds on the fundamental concepts of the *High-Performance Organization Model* authored by David Hannah. The natural, organic, extension to supply chain organization design and optimization is key to achieving a breakthrough in supply chain productivity.

Toyota's standardized work concept has been built on and is presented in *Stress Free™ Work Process Solutions*. This is also applied in improving the work processes along the supply chain.

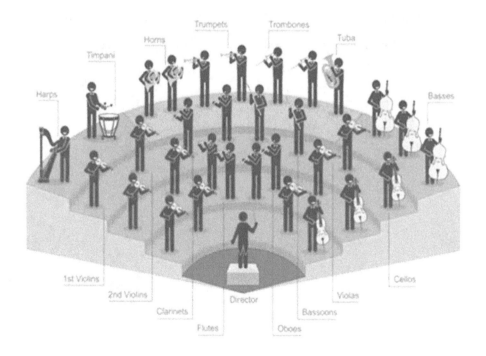

In summary

The supply chain is as complicated as the Boston Symphony Orchestra. The conductor follows the score, the plan, sets the tempo and ensures the right participation of each section of the orchestra.

The conductor has the vision of how the work being performed is to transpire and works with each section of the orchestra to ensure their participation adds to the total score.

It takes much practice to become great and once greatness has been achieved, the great still must practice. The fundamentals still apply.

This book will focus on showing a few key high leverage changes that provide the breakthroughs that will forever envision and empower the way the supply chain leaders and everyone in the supply chain thinks, acts, and performs. It is the score that will lead the organization to greatness.

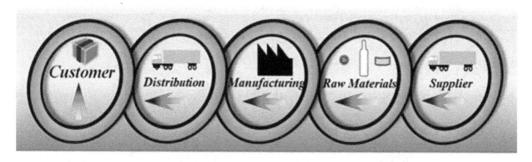

Chapter 1: Supply Chain

In most cases the supply chain is closer to a pile of wriggling fishing worms than a chain. Designed on paper to be very organized and logical, it is implemented based on the conditions and situations of the moment and the contracts and the money available at the time of execution.

The concepts of PULL extrapolated from the Toyota Production System are a far off and almost impossible dream for most businesses. It requires close co-ordination of all the organizations involved in transforming the raw material into the final product. Most organizations face contract and sometime legal issues that present very significant and expensive PULL barriers.

Few organizations are able to make the conversion to PULL a profitable choice.

Business leaders face the daunting challenge of improving their business performance. They need to not only out-perform their competitors but must out-perform their own previous year's performances.

It would seem they are "between a rock and a hard place". They must envision outside the box and it must be a winning solution. They must do this while they continue to make their current system deliver the results committed for the coming year.

The adaptive, vibrant, dynamic leader looks at his business in a new way. This leader sees a new way to improve it in its entirety with a few **short-term** changes and a critical **longer-term** change.

Short-term: Application of Statistically controlled product replenishment.

Longer-Term: Stabilize and Synchronize the production floor and Standardized Work in order to maintain the short-term gains.

The product customer and the consumer at the end of the supply chain define the goals of the supply chain.

This leader looks out to the suppliers of raw materials, the conversion of the raw materials to finished product and the delivery of the finished product to the customer.

This leader can taste the sweet Moscato wine, he can smell apple blossom, and he sees the sparkling mountain stream. A few key changes will dramatically improve the current supply chain performance.

This leader is going to envision his business compatriots to the flowing supply chain.

The supply chain is a complicated network of organizations interacting to move raw materials through to create a finished product. Horizontal organization units interact with vertical support organizations. The flow is horizontal, and the support organizations are vertical.

The flow synchronized supply chain system integrates and co-ordinates the transformation work of everyone along the supply chain. The statistical production flow control provides the cadence for the movement of all materials and finished product. The inventory to support flow is thirty to sixty percent lower and productivity twenty percent higher.

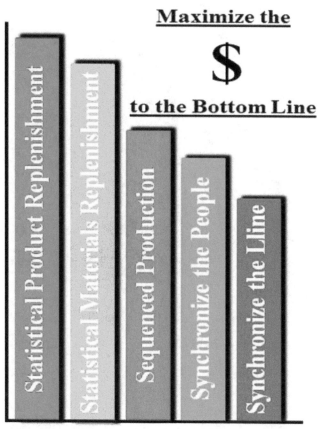

Five key flow improvement factors

Addressing the five flow improvement opportunities is key in taking the supply chain organization along the journey to a flowing supply chain.

These five key flow improvement opportunities are:

1. Production at the rate of product shipment – Statistically Controlled Product replacement.
2. Raw Material flow at the rate of production.
3. Mean Time between production runs and sequence order that match customer shipment.
4. Synchronization of the work.
5. Synchronization of the material transformation.

These five opportunities will be addressed in subsequent chapters.

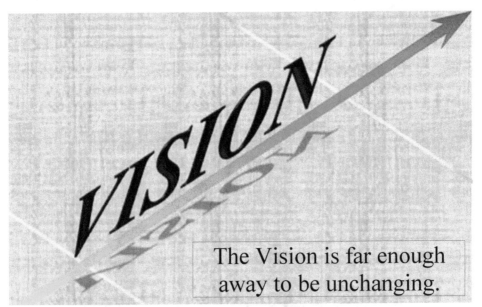

The Vision is far enough away to be unchanging.

Supply Chain Vision

The Supply Chain Vision is a vision that is out three to five years ahead. An example is, "The consumer is always able to buy the product she desires at a price she is willing to pay and every business in the supply chain makes the desired profit margin."

Realizing this vision requires a commitment to a flowing supply chain transformation that starts with a company's top leadership envisioning the benefits of a waste free transformative supply chain, product flow stability improvement, productivity increase.

The leader must declare flow as the supply chain strategy of choice.

This declaration translates into to an organization where everyone will commit themselves to creating statistically controlled flow that will deliver 100% of the winning business results while creating a high-performance workforce that is enabled by an inclusive, team oriented, human focused, learning supply chain culture.

The organization has a long-term vision, and all their people support the vision. What actions must be taken to create and operate the supply chain to make progress toward the vision?

Supply Chain Strategy

The change to something better should not create a business loss. The benefits achieved by making the transition must pay for making the needed changes. It is a business investment decision.

The five key improvement factors provide the basis for deciding on a strategy that generates the money to pay for the improvement.

Strong, hands on, on the floor active leadership is the number one ingredient. *"Give me a strong adaptable leader and I will make a brick productive"* REM.

A requirement is, "A single Supply Chain Leader utilizing statistical replenishment flow control to orchestrate the supply chain across the existing organizational barriers."

Later the supply chain organization will organize as a flow versus a functional organization.

Strategy:

1. Free up cash to pay for the improvement.
2. Understand the main barriers to flow.
3. Enable and empower the people along the supply chain.
4. Create a virtual supply chain organization.

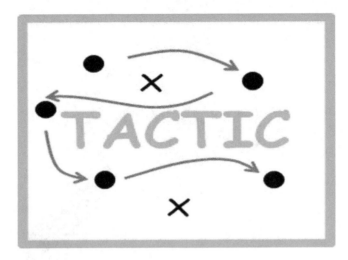

Supply Chain Tactics

1. Apply Statistical Inventory replenishment control to create a steady material and product flow.
2. Utilize Output tracking to synchronize the work along the supply chain.
3. Sequence the production process
4. Implement Daily Management and root cause problem solving to guide daily execution and grow the organization capability and improve long term system stability.

The change is holistic. The transition challenge is felt most directly by the leaders. They must:

- Become aware of the opportunity.
- Learn by doing.
- Do the work to show it is important.
- Envision their organization to the value of operating in flow.

Then they must lead their organization by becoming the coach, becoming supportive, becoming adaptive.

Every leader who has embraced this approach has succeeded. The business experienced almost immediate results and five years later the business results were still skyrocketing.

Supply Chain Measures linkage to Continuous improvement

The ability to measure is critical to developing a leader's supply chain understanding. Each measure must be linked to the "production area" and to the capability needed to ensure the measure is maintained or improved.

A supply chain Measures X matrix like the one shown connects the business measures to specific skills and actions needed to maintain and improve the supply chain. This Matrix is available in the accompanying Stress Free™ *Supply Chain Tools*, excel workbook.

Going counterclockwise the business measures on the left are linked to constituent measures, normally more closely related to the work area. These measures are then related to losses that in turn are linked to specific capability or tools to counter the loss.

This is the stress-free cycle because leaders can find their way to the tools and capabilities that allow them to overcome the waste and losses that hinder them.

Supply Chain Solutions Order

There is an optimum order to all the problems for every supply chain. In most cases the largest barrier to flow is the production planning process.

This can be addressed by.
- Understanding the barriers to flow
- Starting at the dollar income end
- Identifying and prioritizing the barriers to flow.
- Implementing flow countermeasures

It is most interesting that the control signal for production is one of the biggest barriers to flow and flow inventory levels.

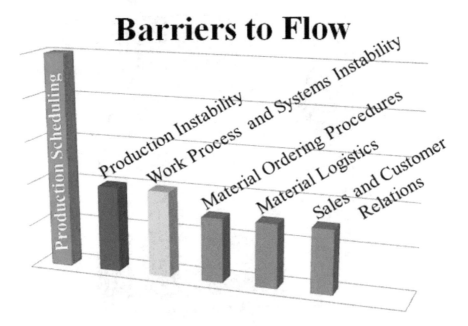

Barrier Countermeasures

- Statistical Replenishment Production

- Root Cause Problem Solving

- Disciplined Daily Management

- Autonomous Maintenance Application

- Planned Maintenance Application

- Supportive Leadership Coaching

- Support functions partnering

- Supplier and Customer Participation

Chapter 1: Tools

Stress Free™ Comprehensive Supply Chain Assessment

Stress Free™ Business Opportunity Statistical Survey

https://www.opcg.biz/stress-free-initial-opportunity-analysis

Supply Chain X Measures Matrix

Chapter 1: Learning points

- The comprehensive supply chain assessment creates common supply chain situation understanding and alignment.

- Statistical inventory product replenishment creates more than 60% of the dollar savings opportunity.

- Raw material inventory reduction organically follows the product inventory reduction.

- The quickest and most beneficial improvement order is to begin with the statistical product inventory control.

Chapter 1: Summary One Point Lesson

Leadership orchestration is required to lead the people in the supply chain in the continuous improvement journey, to a Stress-Free work environment to which these people look forward to coming to work.

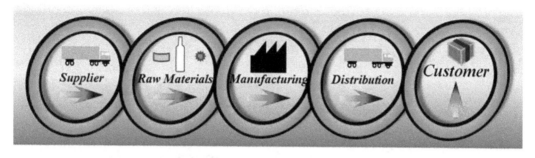

Chapter 2: Supply Chain Leadership

The supply chain leader is the grand orchestra director. More than a mere mortal, this person must have the vision of the flowing supply chain. This person must be able to inspire, to envision, to enable and to empower those compelled by the clarity of the vision that is shared.

In my experience, manufacturing is like the violin section, it delivers the continuous pull for the rest of the orchestra. Each of the organizations directly in the flow of materials and finished product or in support of these organizations are critical in giving the orchestra its depth and resonance.

The high standards of the orchestra are maintained by the lead in each instrument section. It is no less so in a production supply chain. Supply Chain leaders hold the standards because they are masters.

The supply chain is long, complex and has many "leaders". Ideally there would be one owner of the entire supply chain to whom all the other leaders reported. They all should be rewarded for actions that support satisfying what the customer wants and the elimination of losses and waste in the transformation of the raw materials to the finished product.

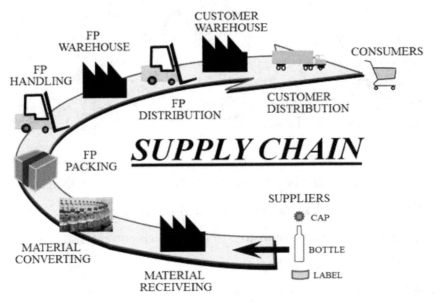

The key ingredient for success is replenishment envisioned leadership. Replenishment based supply chain flow should be thought of as one step short of the Toyota Production System PULL. Every system has its natural, unvarnished, or unpolished flow. It may suffer many significant ills, but it nonetheless has a natural flow. Removing the barriers to this flow will generate substantial, even dramatic, business benefits.

Leadership already knows most of the problems. Every leader can describe what the problem is. In fact, each leader has a slightly different take on the problem, and this often results in disagreement among the leaders.

The Stress Free™ *Comprehensive Supply Chain Assessment* is a half day coach guided qualitative assessment that identifies the *gaps in the current system.*

Leaders are guided to discuss and agree on the rating of key questions that expose the performance gaps in the current supply chain.

The rating discussion among the leaders is critical. It exposes the misalignments and provides the forum to gain the required understanding.

A noninvolved, skilled facilitator is often pivotal in ensuring the assessment makes progress and everyone is heard.

The leaders already know the problems and they often know the improvement action necessary. A key step is to determine the priority of the potential problem countermeasures and any interactions contained in the countermeasures. The prioritization capability increases significantly as the concept of a flowing supply chain becomes clearer.

Specific countermeasure actions are identified, and a preliminary action plan is developed. These are evaluated and a weighted priority pareto is developed.

The result often surprises the leaders and additional alignment discussions occur.

This assessment is often extended to facilitate more in-depth planning for implementing immediate action.

This qualitative supply chain understanding can be enlightening and lead to many very significant improvements but… it must be followed by improvement actions that close the identified gaps.

Quantitative Supply Chain Assessment

Business Opportunity System Survey (BOSS) is the assessment that has statistical teeth. Product by product it identifies the value of the supply chain improvements that can be made by implementing replenishment control.

This survey is done using the past year's daily shipment or sales and the past year's daily inventory. This data is analyzed and statistically determined limits provide the guidance for the production. Production planners determine how to keep the inventory in the green band by authorizing or cancelling prescheduled production.

Product Replenishment Control System

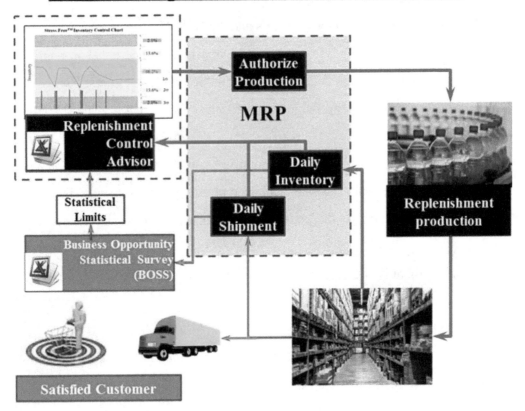

Keeping inventory in the green provides a 99.97% availability rating and an inventory that is normally a dramatic reduction.

The transition shown in the chart resulted in a sixty percent reduction in the average inventory. A key element of this approach is to determine a fixed daily and week to week production pattern. This greatly increases the stability of the transformation processes and results in an increased production capacity.

Over and over this analysis identifies the top ten to twenty products that produce 60% of the profits. It identifies the runner, repeater, and stranger products.

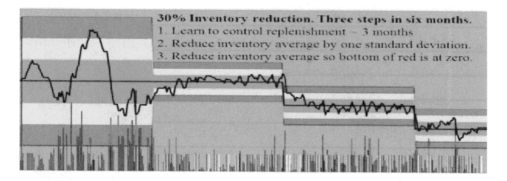

Product Classifications

> **Runner:** a high-volume product produced multiple times a week.
>
> **Repeater:** a moderate volume product produced weekly or biweekly.
>
> **Stranger:** a product that is produced periodically and sometimes
>
>   randomly.

Each classification has a different statistical flow control response. Stranger classified products normally fall into the produce to demand or produce to order category.

A Statistical Flow Control Simulator allows the leadership and key personnel a no risk way to practice the statistical flow control. This prepares them to utilize the Statistical Flow Control Advisor* to guide them in controlling their MRP (**M**aterial **R**equirements **P**lanning) system.

*OPCG provides initial analysis and prepares the statistical control simulator and final advisor.

Supply Chain Oriented Leadership perspective.

Think of the supply chain as a multi-lane, super-highway with convenient on and off ramps for all normal traffic and for the support and emergency support vehicles.

The materials flow smoothly and effortlessly along this super-highway and merge to form the desired product. The product consistently reaches its destination, at the desired time, the desired quantity, and the desired quality.

The many drivers along this super-highway do their work smoothly and effortlessly. They enjoy the work that they have helped to optimize. They feel great about helping every one of their constituents. They know and work closely with their immediate customer and they understand their contribution to the final product and paying customer.

The supply chain leaders, the highway patrollers and maintenance crews, keep the lanes of the highway well maintained and the flow at the required safe speed with everyone in their proper lane. They ensure the support organizations are properly positioned and know when to come on and where to get off.

There are no toll booths, no unexpected reduction in the number of lanes, no unplanned increase in the flow of traffic. The emergency lane is always open but seldom needed.

The designers of this supper highway engaged the patrollers, the maintenance crews, the drivers, the supporters, the material senders, and the product receivers in the original supply chain organizational design.

It becomes the supply chain leader's role to:

- Maintain the supply chain in a long-term operational condition,
- Become supply chain oriented versus technically oriented.
- Become an Adaptive, hands on coaches.
- Communicate broadly to everyone along the supply chain.

Chapter 2: Tools:

Qualitative Assessments –

Stress FreeTM Comprehensive Supply Chain Assessment

Quantitative Assessment –

Business Opportunity Statistical Survey (BOSS)

Statistical Replenishment Simulator

Statistical Replenishment Control Advisor

Chapter 2: Learning Points:

- Leadership defines, creates, and maintains the supply chain culture.

- A product by product statistical control replenishment system achieves dramatic business results.

- A master plan to improve the supply chain organization typically has the following elements.

I.	Supply Chain Organization Design Evaluation
II.	Information System Improvement
III.	Daily Management Enhancement
IV.	Work Process Improvement
V.	Production Equipment and System Improvement
VI.	Logistics Systems Improvement
VII.	Supplier Engagement
VIII.	Support area engagement
IX.	Customer engagement

Chapter 2: One Flow One Point Lesson

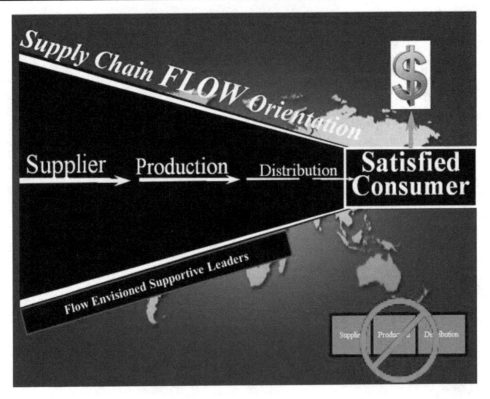

The winning mutually beneficial relationships of a flow-oriented leadership and organization versus a compartmentalized one.

Chapter 3: Statistical Product Replenishment

Toyota has inspired the world with the concept of PULL. The product pulled by the customer is immediately replaced and available for the next customer. Inventories along the supply chain are kept very low. Waste and supply chain cost is targeted for elimination.

The Toyota production system evolved over time and became the fabric of how the Toyota supply chains are established and managed today.

Toyota leadership was requested to provide guidance in the application of PULL to a Consumer Goods Company (CPG).

The Toyota leadership warned that PULL would not be possible for a system running at the high production speeds and the complexities both contractually and physically of the logistical systems associated with one of the simplest products being produced.

The CGP Company leadership insisted we would learn how to do it. At the seventh request for TPS guidance, Toyota relented and assigned a TPS coach to guide the PULL learning effort.

With direct coaching by a Toyota TPS coach, the approach was applied to a consumer goods production system. A very thorough and patient TPS coach[1] guided us in how to implement the PULL concept.

The project was declared a learning success. We learned the application of Standardized Work[2] and successfully applied it. We enhanced the root cause problem solving process. We learned about Kanban[3] and its application. We learned about Takt[4] time and how it might be applied to the product flow. The production improvements were many and the project was a success.

Though many of the TPS concepts were deemed very useful and became part of the fabric of the production System, PULL as a system did not happen.

The daily shipment data analysis indicated that by determining production in a similar manner as the concept of Kanban and by leveling the quantity of each production run, inventory could be lowered while at the same time product availability could be maintained or increased.

Learning how to set the statistically determined average ideal daily inventory was a breakthrough.

Subsequent production site trials verified that controlling the flow of production in this new manner supported lower inventory levels while achieving higher customer shipment results.

The statistical analysis and method of determining the ideal inventory level and the width of the control limits were field verified multiple times. In every case, there was improvement. In a few cases the fact that inventory was being held too low was a learning surprise.

Though successfully demonstrated in several other sites, the effort met huge resistance because it was not integrated into the Enterprise Resource Planning (ERP) system, and it does not utilize product forecasting for production control

Current Situation:

The technique is based synchronizing to the daily shipment, daily use, or daily sales history. This history is used to determine the standard deviation of the ideal inventory that when properly replenished will meet all use variation in the future.

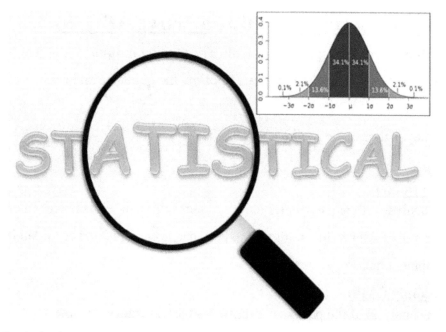

Statistical Inventory replenishment works across many businesses, consumer goods production, spare parts optimization, internet sales and distribution, filling operations, and intermediate goods production. All can benefit.

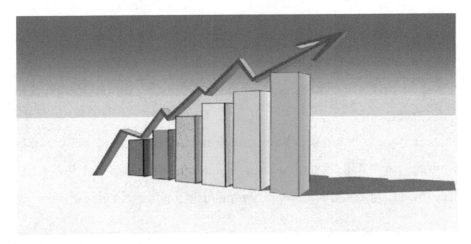

Business Opportunity Statistical Survey (BOSS)

BOSS is a very appropriate acronym. The guiding information it generates has proved to be on target and accurate. Follow the guidance and achieve the desired results.

The key is to provide it with accurate data.

The Data

The Customer

The analysis of the past year's daily product shipment tells the customer behavior story. All product logistical experience is captured and recorded in the daily shipment data.

The Supply Chain

The analysis of the past year's daily product inventory tells the supply chain response behavior and capability. It is the story of the good, the bad and the ugly.

Inventory is the result of the current system to customer demand mismatch and system transformation and transportation constraints and problems.

A perfect match and no transformation problems would result in no inventory need.

The improvement opportunity is to match the production quantity as close as possible to customer demand and to account for the delays of transport.

The Analysis

The daily shipment and daily inventory history provide the basis to identify the ideal inventory. The resulting inventory average and the standard deviation of the ideal inventory, based on the system defined replenishment pitch, are used to set the inventory response boundaries. These boundaries* become the main guides utilized to manage the flow of production.

BOSS Input

Daily shipment for the past year.
Daily inventory for the past year.
Production/throughput/supply chain pitch

BOSS output

The output is by specific product ship unit.
Target product inventory average.
Reaction limit values.
Number of required production runs.

Statistical Replenishment Control Practice

Change requires practice. The work is different. It is more controlled and precise.

The transition to statistical control is facilitated by the use of a Replenishment Control Simulator. This simulator utilizes the same data used for the BOSS analysis. It provides both numerical and visual feedback to the actions authorized by the product replenishment planner.

With the simulator, the product replenishment planner practices maintaining the target average product inventory on last year's shipment data. With this real data, day to day, simulation allows the planner to experience the full range of work processes required to do the work in the highly variable real-world product shipment environment.

The production system issues can be surfaced and discussed. The potential problems can be addressed prior to the transition to actual product replenishment control.

●OPCG provides initial analysis and prepares the statistical control boundaries

Statistical Replenishment Control Advisor.

The product shipment and the product inventory value for the previous day is extracted from the existing system Material Requirements Planning (MRP)[*5] being utilized and put into the Statistical Control Advisor.

The Advisor shows the trend of the Inventory and visually enables a decision on whether a production run is needed at the specified pitch. The human replenishment planner evaluates the situation and decides whether to authorize the production as planned. The authorization choice is then put into the MRP system.

The controller is designed to flag the incoming day's data as red, yellow, or green. The replenishment planner is coached to review and address the products in the red band first. Then address those in the yellow band. The products in the green band are within the target limits and do not need immediate attention.

The Statistical Replenishment Control Advisor[*6] provides dramatic improvement by stabilizing the entire supply chain, and product flow to the customer.

The control advisor utilizes.

- Daily Sales or shipment to better understand the Customer.
- Daily Inventory to better understand supply chain situation.
- Utilizes the variability generated by creating the ideal inventory level to establish the red, yellow, and green control limits.
- Biases the Ideal Inventory Average up by; + 3 sigma to ensure a 99.97% probability that inventory is available to meet the ordered shipment. This bias can be more conservative than six sigma as may be the case for the medical businesses.

Product Replenishment Control System

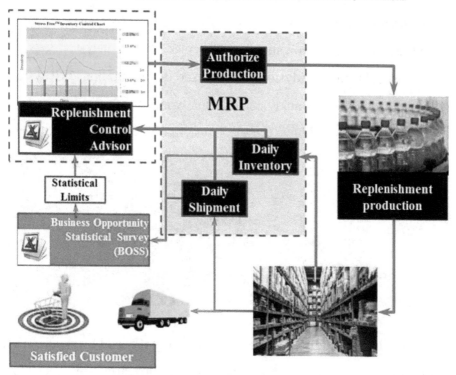

Product Replenishment Control System

This is currently an advisory system that provides the production planner daily shipment and inventory situation. The planner must still look forward to determine the required short-term upcoming production need and compare it to the preplanned scheduled production. This preplanned production is based on meeting the previous year's average shipment.

The planner then follows the simple replenishment authorization rule of only authorizing the production if the inventory level is into the lower yellow control band. If the inventory level is in the green and the forward look does not require additional finished product, the upcoming production is called off.

This approach coupled with sequencing of the production process to an optimum sequence and a repeating fixed daily production pattern creates supply chain wide stability.

Sequencing: Establishing a fixed production order and quantity for each day of the week.

The planner utilizes the power of the MRP system to communicate, to authorize the actual production and to order replacement material.

The Statistical Control Advisor can be linked directly to the MRP system. This allows the planner to quickly and efficiently make the required production authorization decisions and have the MRP system operate in its normal way.

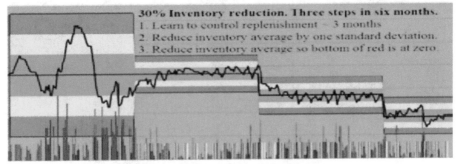

30% Inventory reduction. Three steps in six months.
1. Learn to control replenishment – 3 months
2. Reduce inventory average by one standard deviation.
3. Reduce inventory average so bottom of red is at zero.

The result

The variability of production quantities is greatly reduced by producing to the defined pitch and quantity. In all cases to date, this variability, once reduced allows for the reduction of the average inventory target by as much as sixty percent.

The wide green, yellow, and red bars show on the left side, show the operating variability in a normal production system. The application of the Stress Free™ Statistical Product Replenishment Control results in the greatly reduced red, yellow, and green control bands. This leads to the inventory reduction opportunity.

The first step is to demonstrate this tighter control to maintain the current inventory. This is the first reduction in the green, yellow and red bars. Then a series of reductions can be taken until the bottom of the lower red band reaches zero.

The replenishment planner, utilizing the control advisor can maintain all product SKU's in the same fashion.

Chapter 3: Tools

Stress Free™ Statistical Product Replenishment Simulator

Stress Free™ Statistical Product Replenishment Advisor

Chapter 3: Learning Points

- Last year's daily product shipment and its standard deviation defines the minimum inventory.

- Control bands based on the standard deviation provide useful guidance for current production decisions.

- A set production pattern with a specific constant production volume target provides flow and work stability.

- Significant inventory reduction is the result of staying in control and in the green.

- Reduction is controlled by the production replenishment planner.

[1] TPS coach Satoko Watanabe
Now Head of the:
Center of Manufacturing Excellence
P.O. Box 1848
University, MS 38677

[2] Standardized Work has three main components:
1. Adheres to takt time.
2. Standard work-in-process (WIP) is specified.
3. Defined sequence of operations for a single person.

[3] Kanban: A Japanese term meaning "signal". It is used in Just in Time (JIT) manufacturing systems. It authorizes a cycle of replenishment for production and materials.

[4] Takt is the German word for the baton that an orchestra conductor uses to regulate the tempo of the music. It is a measurable "beat time," "rate time" or "heartbeat." Takt time is the rate at which a finished product needs to be completed in order to meet customer demand.

[5] Material requirements planning (MRP); a production planning, scheduling, and inventory control system used to manage manufacturing processes.

[6] Statistical Replenishment Control Advisor is a product supplied by Optimum Performance Consulting Group (OPCG). OPCG also has the Replenishment Control Simulator.

Chapter 3: One Point Lesson

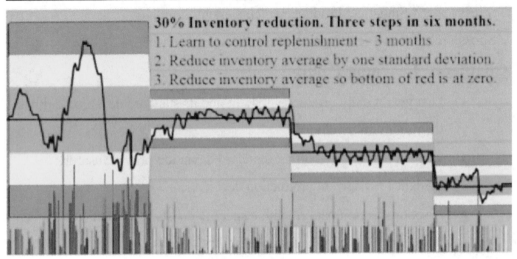

30% Inventory reduction. Three steps in six months.
1. Learn to control replenishment – 3 months
2. Reduce inventory average by one standard deviation.
3. Reduce inventory average so bottom of red is at zero.

Key aspects: Production Quantity prescheduled, fixed frequency Control bands with reaction guidance.

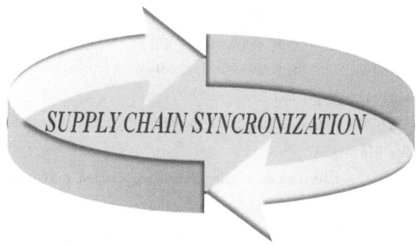

Chapter 4: Supply Chain Synchronization

Supply Chain Optimization focuses on the improvement of existing product delivery systems. The best approach would be to design and implement cost efficient supply chains. For many reasons, from taxes to previously existing systems, most supply chains get designed and implemented in a less than optimum configuration.

Good supply chains slowly get buried under layers of change and incident triggered practices. These practices at one time were improvements or defensive in nature. Later they became imbedded in the normal practices of the delivery stream.

Just reviewing the flow of materials, information and observing the transformation and transportation taking place along the supply chain will yield improvement opportunities. Applying a series of simple tools and practices in a strategic way will yield fast and lasting continuous improvement.

The cultural change in the organization is critical to superior results over time.

Supply chain synchronization means that time, material quantity and flow get coordinated along the entire supply chain. This includes the support from engineering, quality, maintenance, logistical, sales, finance personnel and many other groups.

Leadership must cross the organizational lines. The organization needs to change to a supply chain focus versus a functional focus. Synchronization requires each functional organization to change its work process to match the timing required by the product flow drumbeat.

All supply chain participants must be "dancing" together to whatever music the supply chain plays.

Statistically controlled product replenishment requires that the **entire supply chain follows the metronome established by the** practice of producing in the green with fixed volume sequenced production authorization. The constant rhythm of flow moves up the supply chain to the supplier and his supplier. It moves out to the supporting organizations and must be reflected in the monitoring and control systems. The material flow to the final product is the backbone of the supply chain. The information and communication systems are very critical in synchronizing the supply chain.

Material Flow

Material flow will follow the behavior of the statistically controlled product replenishment. The timing and quantity of material movement will match the statistically controlled production process.

Follow the raw materials from the product production line to the material supplier. Note every time the material gets touched by another person. Each touch by another person represents a flow boundary. The more touches, the more boundaries, and the more losses.

> **Touch:** Any time the product or packaging material gets handled. This occurs during material receiving, materials staging, and materials delivery to the line, line feeding and when the finished product is moved to the warehouse and finally loaded into a truck.
> **Boundary:** Raw Material Truck loading, Transport, Raw Material unloading, Delivery of Raw material to production line, finished product removal, finished product storage, finished product loading into truck, delivery to customer.

The material flow boundaries must be seamless, and they must be synchronized in time and quantity to the authorized statistically controlled production replenishment.

Boundaries are most often the points where rework and waste are generated. Resolving the problems at each boundary results in the reduction of supply chain operating cost.

Stabilizing incoming material flow implies replenishment control of raw materials from the raw material suppliers. The coordination to ensure material flow stability will require new relationships with key raw material suppliers. Key raw material suppliers become part of the improvement team. These suppliers will experience a significant reduction in their trapped cash. Their hands-on personnel interact with your material receiving hands-on personnel.

Output tracking is implemented on the supplier end and material receiving tracking is implemented on the receiving end.

If the logistics between the supplier and the customer includes a transport company, the transport leadership and schedulers are brought into the supply chain stability work.

Stabilizing line delivery requires the flow from the material delivery truck to the line to be studied. The focus is on the materials associated with key high-volume SKU's. Understanding and reducing the number of touches will yield significant productivity improvements. The quantity and the frequency of delivery to the line is now determined and authorized by the replenishment planner. The variability of this delivery is monitored by the use of delivery tracking

Information and communication system

These systems are critical in the longer term in establishing the supply chain synchronization. In the short term, they actually are part of the problem! They are replete with problems that must be fixed. These problems self-identify with the use of output tracking.

Information Systems

Data may not be kept in the granularity that supports longer term improvement. Daily shipment and daily data may not be linked or even measured at the same time. The data is often stored and then forgotten.

Communication systems

These systems are used most often as the means to call for help versus a means to prevent a problem. Communication must utilize the visual, the audible, the computer, and internet avenues to be fully effective. Wireless communication now enables equipment to communicate directly to people, no matter where they may be located.

This area is rapidly changing, and new opportunities are constantly surfacing.

Boundary hourly throughput tracking is the primary way to identify the material and product flow issues. Clarity on how to measure the flow across the boundary and then the flow through the system to the next boundary is critical to establishing and maintaining flow.

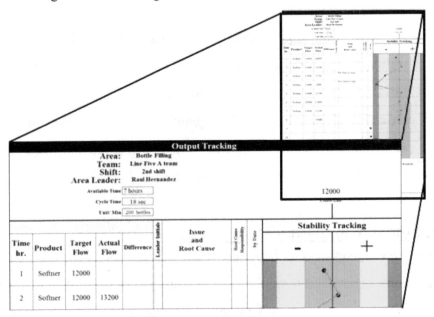

Example output tracking sheet.

Output Tracking.

The tracking sheet is the same at every boundary. It may be titled differently but the tracking is normally done in the same manner.

Output tracking, Input Tracking, Throughput Tracking are all basically the same. "A rose is a rose by any other name," and so it is with Output tracking.

- **Supplier Shipment Output tracking**,

  The supplier leadership is trained by on replenishment control.

  Output tracking is implemented on the supplier end and material receiving tracking is implemented on the receiving end. If the logistics between the supplier and the material receiving includes a transport company, the transport leadership and schedulers are brought into the supply chain stability work.

- **Material receiving Input tracking**,

  Stabilizing incoming material flow implies replenishment control of raw materials from the raw material suppliers. Key raw material suppliers become part of the improvement team. Their hands-on personnel interact with the material receiving hands on personnel.

- **Line delivery Output Tracking**

  Stabilizing Line Delivery, defined as movement of raw material from storage area to the production line, requires the flow from the material delivery truck be studied. The focus is on the materials associated with key high-volume SKU's. Understanding and reducing the number of touches will yield significant productivity improvements. The quantity and the frequency of delivery to the line is now determined and authorized by the replenishment planner. The variability of this delivery is monitored by the use of delivery tracking.

- **Production Output Tracking**,

  Stabilizing finished product replenishment is the focus for the production area. Work on the reliability of the equipment yields increased production area stability. The average and 3 sigma determine the inventory that must be held due to transformation instability.

- **Finished Product Loading Tracking,**
  Stabilizing finished product loading focuses on how effectively the loading process is managed.

  "Any loader, any SKU, any truck" is an approach that will reduce the miles traveled by the forklift truck. Truck loading variability, as measured by tracking each truck load time, highlights the variability of loading. Reduction of the variability normally results in reduced loading time.

- **Customer Receiving Tracking**
  Stabilizing finished product delivery to key customers is a first step in engaging our paying customer. This should be done in mutual learning mode. It requires the engagement of top customer leadership. This is a big deal and only customers known to be willing learners should be approached. They need to become partners in the flow improvement team.

True customer need understanding must be the driving factor for every boundary in the organization. This need should be visually recognizable and a miss in need delivery should immediately raise a red flag.

Once the customer is engaged, they can utilize receiving tracking and communicate any issues back to their supplier

Customer receiving tracking will become the ultimate in determining the customer service achieved by the Supply Chain.

Chapter 3 Summary

Variations based on time span or quantity may make the input or output tracking appear slightly different, but they all serve to measure the performance variability in a specific area. Every boundary is measured, and the variability problems are addressed as the continuous improvement focus for the area.

The critical element is that it:

- is managed by the working shift,
- has a specific flow target for each shift,
- follows the production pitch drumbeat,
- it is visually graphed,
- it is deployed on every boundary.

Chapter 4: Tools

Output tracking

Stress Free™ Manufacturing Solutions

Chapter 4: Learning Points

1. Shipment frequency and volume is the supply chain synchronization signal.

2. Synchronization is managed with output tracking.

3. Output tracking identifies the boundary issues or irregularities impeding flow.

4. Issue countermeasures are usually rapidly implemented.

5. Today's communication technology can be leveraged to accelerate synchronization.

Chapter 4: One Point Lesson

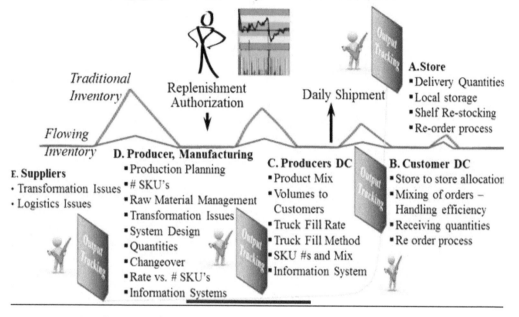

Output Tracking
Supply Chain Synchronization

- Implementation along the entire supply chain is key.

- The replenishment planner authorizes replenishment.

- The replenishment value is tracked on the output tracking sheet.

- The output tracking sheet is a daily management tool.

Chapter 5: Supply Chain Organization

Flowing Supply Chain Organization

The chief operation officer must envision the top thought leaders to the benefits of moving from organization driven to a capability building Supply Chain focused approach. They can then invest their time and energy in developing an organizational design and structure to deliver the transformation and enable and energize the organization to build capability to move the supply chain organization forward.

Supply Chain improvement deployments leverage the scale and power across the entire business.

The supply chain organization can expect 3-5% margin point improvement, a 40% reduction in inventory and a > 20% improvement in productivity, significant reductions in both quality defects, and serious safety incidents, increased throughput, and customer service improvement.

The resulting high-performance culture will ensure the retention of the champions in the organization

Achieving a waste / loss free, long term sustainable supply chain from raw materials to a consumer preferred product will provide a breakthrough. The waste / loss free approach is more profitable, agile, and flexible. This winning way can only be achieved when all resources work harmoniously together against common goals and a clear vision.

With hands on experience, the leading champions have the means of bringing the flowing supply chain vision to life. They are able to convey how they see it improve the lives of everyone and improve the long-term results of the business.

The vision becomes more than a slogan. It becomes the rallying point for all the people. It encompasses the entire organization. It becomes the Supply Chain Vision.

Every champion goes through the emotional high when they experience firsthand the power of the Stress Free™ Approach.

To create value, the customer need must be immediately addressed, waste must be reduced, throughput must match the need, and new initiatives must be brought efficiently on-line.

Each supply chain organization needs an improvement coach. This improvement coach provides an external perspective. The coach has a broad level of hands on capability. He/she sees many businesses. These coaches normally coach four to five sites.

An external coach provides the supply chain a different set of eyes. This external perspective allows the losses that have become a part of the supply chain to be identified and eliminated.

A series of workshops provide the hands-on experience to the key supply chain improvement leaders. These leaders along with the supply chain coach develop a comprehensive, pay as you go deployment plan.

The vision is to start from any situation and to take the journey to waste / loss free with all the people involved.

This never-ending drive to eliminate waste and loss will deliver breakthrough business results.

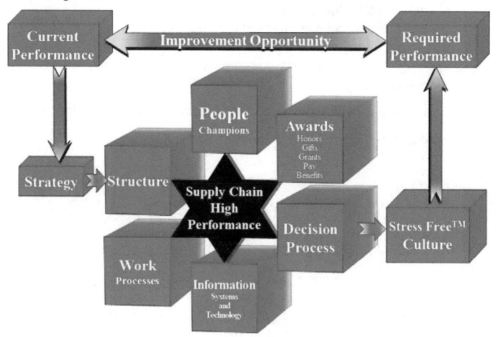

<u>Supply Chain Organizational Assessment</u>

Leaders need a clear understanding of the entire supply chain situation. They need alignment to the most critical maintenance and improvement actions to take.

A comprehensive supply chain assessment on at least a yearly basis provides the leaders an aligned understanding of the supply chain situation. The assessment effort need only be in as much depth as required to get leadership consensus on the continuing supply chain improvement effort.

The high priority improvement efforts should surface during the alignment to the assessment question answers.

The assessment questions need to evaluate the condition of the supply chain in:

- leadership and the business situation reward
- organization
- information and material flow processes
- material transformation production processes
- finished product handling and distribution processes
- product to the shelf processes
- support and new initiative processes
- rewards and people

The assessment provides the supply chain leaders a clear understanding of the gaps and guides them to make the improvement selection.

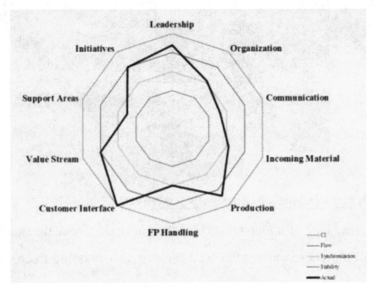

Visual output of a Supply Chain Assessment

The supply chain champion leaders gain and embrace the concept of a flowing supply chain and the assessment highlights the gaps. Enrolling the supply chain organizations as a whole is critical. All parts of the supply chain need to move forward together. Not all at the same speed but each area must identify their role and part in the continuous improvement journey.

"When the mind functions, the muscles respond automatically, the heart pumps at the rate dictated by the muscle and the muscle responds based on the load."

The supply chain functions in a similar manner. The relationships between its various parts have been established on the basis of need. These relationships and the established organizational boundaries require examination. The boundaries are normally where waste is created.

Today communication technology allows every business result champion to know immediately when a sold unit goes out the door.

There are no primary, no support, and no secondary functions. There are only supply chain result Champions.

Supply Chain Result Champions

Supply Chain Enterprise Champions

- Supply Chain Leader; site, region, country, international.
- They are the orchestra conductors that keep all the members synchronized.

Customer Facing Champions

- The Site Manager
- The Sales Managers
- The Financial Contract Managers
- The Logistics Managers

These are the champions that must envision the customer to the impact that a steady statistically controlled flow closely meeting his consumer demand is an opportunity to free the cash trapped in the back room and on the shelf.

Product Design and Delivery Champions

- Product Development
- Process Development
- Engineering and Initiative Delivery

These folks deliver the new or improved product that seamlessly comes into the stably flowing supply chain.

Supply Chain Organization Operations Champions

- Site Manager
- HR manager
- Safety Manager
- Quality Manager
- Engineering Manager
- Support office managers

These are the people that ensure the supply chain organization has the quantity and quality of people and functions as intended. They monitor and ensure standards are followed and continuously improved.

They support and coach the organization in maintaining high safety standards and behaviors.

Raw Material Flow Champions

- Material receivers
- Line material deliverers

Material touches are minimized, and the quantities taken to the line exactly match that authorized to maintain flow. The precision of material delivery quantity reaches the point that line clearance is less than a kitchen trash bag.

Transformation Champions

- Line operators
- Maintenance personnel
- E&I and Electrical personnel
- Team Leaders
- Line Leaders
- Department Managers
- Operations Managers
- Site Manager

These are the people that ensure the material is safely transformed to the desired product at the specified quality and authorized quantity. They keep the equipment at target performance and operating speeds.

These folks become expert root cause problem solvers and continuous improvement champions.

They make work easy to do, mistake proof and are always improving its efficiency.

Those directly handling the materials and interfacing with the equipment know their leaders are capable and ready to support them as needed.

Problem solving resources are visually signaled when needed and resolve problems immediately.

Finished Product Flow Champions

- Finished product to Warehouse handler
- Warehouse Forklift driver or other order loaders
- Truck Scheduler
- Customer order verifier

The product flows to replenish the inventory that has been loaded onto the out bound delivery trucks or into containers for rail or ship transport.

The customer orders are accurately filled and the finished product flows to the customer on time, on quantity and quality.

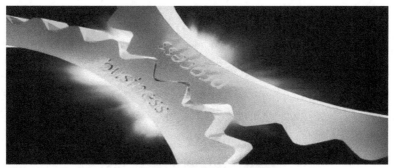

<u>Supply Chain Optimum Organization Design</u>

The Comprehensive Supply Chain Assessment questions and the evaluation of answers are a reflection of the Supply Chain Organization Design and the current social and work culture. The high-performance supply chain design is not monolithic nor structurally rigid. It needs to be organic and adaptive. It must keep up with technology and rapidly changing social culture.

Understanding the situation is critical. The supply chain crosses multiple organizations and though they may be in constant contact, leaders seldom take the time to gain a common alignment on the issues before them.

A comprehensive supply chain assessment utilizing key questions that expose the performance gaps in the current supply chain form the basis for a common alignment. Specific countermeasure actions are identified, and a preliminary action plan is developed. This assessment is often extended to facilitate more in-depth planning for implementing immediate action.

The opportunity may be immediately known, and a specific action taken.

Organizational Design (1977) by Jay R. Galbraith led to the codification of P&G's High-Performance Work System HPWS organizational model. This was further developed by David P. Hanna at P&G and later published by him externally as *Designing Organizations for High Performance* (1988).

Neither Galbraith nor Hanna was thinking in supply chain organization terms. Neither of them had the information technology of today. The basics they described have not changed but organization design implementation then, is now similar to baking a cake in a wood burning versus a microwave convection oven.

Now in 2020 thirty years later it is time to think in terms of the High-Performance Flowing Supply Chain Organization.

The organization must empower all people and make them champions. This is accomplished by implementing the High-Performance Supply Chain Work System Design.

The foundation of this design is the foundational principle of treating everyone as you wish to be treated.

How the supply chain works, how it grows, how results are measured and delivered and the strategy that holds it all together leads to an organization of champions.

Each element must be designed to produce the desired output at each supply chain boundary. Previous chapters highlighted the strategy of flow and flow implies an organization with as few boundaries as possible.

High Performance Supply Chain Work System

Supply Chain Structure

The flow-based supply chain structure implies the material transformation process management is the primary element that defines how to organize to win.

Supply Chain Tasks

The supply chain tasks are designed to operate the transformation processes and maintain the steady flow that matches the statistically controlled replenishment authorization.

The supply chain daily management system can be tuned to support replenishment.

Supply Chain Information System

The information system continuous to be one of the most rapidly evolving opportunities. Given the right application of technology, every individual can be synchronized to the replenishment authorization signal. Every task can be adjusted and tuned to this signal. Non-value-added work can be minimized or eliminated. Required work can be scheduled appropriately.

Supply Chain Rewards

The Supply Chain Champions deliver the business objectives and goal. The reward system must recognize this with social recognition as well as monetary rewards.

Supply Chain Decision Making

The supply Chain decision making must be designed to occur at the lowest level possible and the capability to make good decisions trained to each individual.

Supply Chain Champions

Continuously developing and renewing people's skills creates an organization of winners. Organization winners challenge and reinforce each other as they deliver superior results.

Supply Chain Optimum Organization Performance

The creation of winners occurs in a learning organization supported by strong adaptive capable leaders. Leadership creates the culture and the people he or she supports create the superior business results that allow the business to win in the marketplace.

Tools for Chapter 5:

1. Comprehensive Supply Chain Assessment
2. Leadership Capability Assessment
3. Organization Assessment
4. Vision Creation Guide
5. Master Plan Creation
6. Measures to Capability Matrix\

Chapter 5: Learning Points

- The Principle of Inclusion is the supply chain foundation.
- Team-work concepts make it happen.
- A learning supply chain environment fuels improvement.
- Delivering results is critical.
- Supply Chain Flow maximizes results

Chapter 5: One Point Lesson - Supply Chain Operation

- Leadership behavior establishes the culture.
- Supply chain organizational design, maintains and continuously keeps it current.

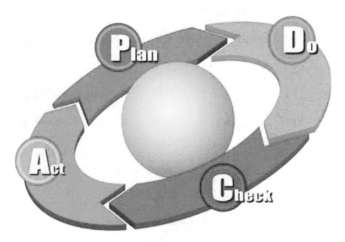

Chapter 6: Supply Chain Daily Management

Supply Chain Daily Management (SCDM) provides the supply chain leader the ability to understand, manage and synchronize the flow of information, and materials to ensure resources are functioning optimally. It is a systemic means of ensuring high supply chain performance.

The elements of daily management all come together to ensure the future twenty-four hours will perform flawlessly. The previous day supply chain performance is examined, countermeasures correcting any shortcomings are assured, the next twenty-four hours are anticipated, and any continuous improvement actions are reviewed.

The supply chain daily management is the last level of the daily management chain that begins at the production line level and rapidly progresses to the supply chain level. The daily management at the line level is based on the operating shift line team and may occur seven days a week. At the supply chain level and for the support functions it occurs once a day, and in most cases only five days a week.

The primary purpose of every level of the daily management system is to organize and apply the existing resources to ensure an, as planned, loss free performance, for the next twenty-four hours and beyond.

Daily supply chain daily management does deploy the company's strategy and goals. It does share the appropriate KPI's at each level of application. It is based on teamwork.

The daily management activity is not about deploying from the top but requesting resources as needed from the production line operating level all the way to the supply chain level.

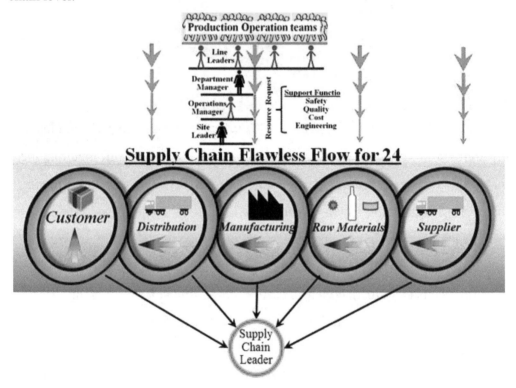

The diagram, Supply Chain Flawless Flow for 24, depicts the operating teams at the top. Daily Management is designed to ensure the operating teams have the skills and resources to keep the production lines running at the required statistically controlled product replenishment pace.

Output tracking provides the details of any issues and is brought to the daily direction setting meeting when an issue resolution requires resources outside of the line operating team.

Daily Management of course is also an excellent organizational structure to deploy the business goals as well as to capture operational performance. This is visually displayed on the daily meeting board but needs little discussion unless a performance gap exists. On a yearly or major event basis the business strategy and Key Performance Indicators (KPI's) are deployed via the daily management interaction.

Coordinating the Supply Chain Daily Management system requires one to understand the vertical resource management occurring on the vertical axis from the production line to the supply chain and horizontally along the supply chain back from the customer to the raw material supplier.

The supply chain organization has many interacting, interdependent elements. The focus in this book will intentionally be kept to the vertical flow from the production line to the supply chain organization level and from the customer to the supplier. This approach will show the materials handling, manufacturing, and product distribution practice of the daily management. The supporting organizations play partnership roles in ensuring the materials get transformed into the desired product.

Daily Management is a foundation system for all organizations. It must flow from the transformation work point to those at each level that control resources. In each case, resource allocation decisions need to occur at the lowest level possible.

Resource allocation is a key issue for all organizations. A robust daily management system reaching across the entire organization that strategically optimizes resource utilization, accelerates the business results and at the same time creates the on the floor culture critical for continuous improvement.

There are some basic requirements that need up front definition and clarification.

1. **On the floor:** the location where material or data transformation takes place in the journey to becoming a product.
2. **Line Operating Teams:** A team of people that work on the same line and set of equipment on a regular basis.
3. **Operating team leader:** One team leader for each line

operating team. The operating team leader maintains the line output tracking chart. Any hourly, off target production is a defect to be evaluated. This implies immediate problem-solving capability.

The operating team leader determines if the operating team has the appropriate problem-solving skills. If so, the team solves the problem immediately. If not, the team leader makes a call for support to the line team leader.

4. **Line leader:** has 24-hour line performance accountability.
   This person manages the teams assigned to one line. This person may manage multiple lines. A key responsibility is to monitor the hour by hour line output tracking.

A miss on quantity, time or quality is a defect to which corrective resource effort must be allocated.

The line team leader may have the supportive problem-solving coaching skills and helps resolve the problem. If not, the line leader will take this resource need request to the department daily morning meeting.

By addressing the defects as they arise, the production transformation process is continually improved.

At the beginning of the journey to supply chain flow, the department and operations managers must step in to lead this problem resolution. At each intervention they must strive to develop their people into becoming master problem solvers.

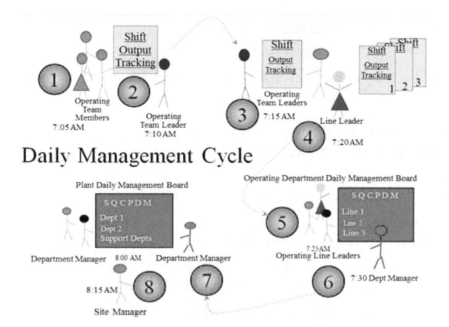

Daily Management Cycle

Daily Management – from the line to the supply chain
Production Area Daily Management

Let's look first at how the daily management flows from the floor to the site leadership and then let's look at how Leadership interacts with this flow.

The night shift operating team, at the end of their shift meets briefly at their operation line team board to review their output performance. The operating team leader joins them to better understand the overall shift performance.

Then the operating line team leader meets with his on-coming counterpart and shares the off going shift's output tracking chart. The line leader joins them and reviews the last three shift output tracking sheets. Together they look ahead to the actions necessary to make the next twenty-four hours loss free.

The line leader takes this information and any resource needs to the department meeting where she meets with the other line leaders and department manager. Resource needs are identified and allocated if available.

The department manager takes his needs and the needs of his line leaders to the site managers meeting. There he and the other managers share their situation. The resources still needed get evaluated and help allocated on a priority basis.

Site Manager and Leadership Team Tour

The site manager follows a set, weekly and monthly operations interaction, routine. Each day of the week, the site manager will tour a specific path through the site.

Day one: Outbound truck loading to finished product storage.
On tour: Site manager, Department Manager, Quality Manager

Day two: Finished product storage to production line one,
On tour: Site Manager, Receiving Manager, Safety manager

Day three: Finished product storage to production line four,
On tour: Site Manager, Quality Manager, Engineering Manager

Day four: Line delivery to the line back to raw material receiving.
On tour: Site Manager, Department Manager, Finance Manager

Day Five: Support areas, Maintenance areas, Environment,
        Safety, Health, Quality, Engineering, Finance, Office area.
On tour: Site Manager, Department Manager, Logistics manager.

Each day's tour begins immediately after the Daily Morning DMS meetings. The weekly tour cycle is repeated every week. A leadership designate is appointed to make the tour even when the site manager is absent. The tour focus is on the adherence to standards, safety and cleanliness, everything in its place and in recognizing and giving feedback to the good work that is being done.

Supply Chain Daily Management

The Supply Chain Daily Management meeting occurs at the same time every day. Each of the supply chain organizations hold similar meetings to what has been described above and then these supply chain organization leaders attend the final daily supply chain management meeting.

This meeting is held at high noon every day. It is a short meeting unless there is a high-level issue that one of the organizations has not been able to resolve.

Boundary flow Issues for the next twenty-four hours are resolved. Longer term continuous improvements are reviewed.

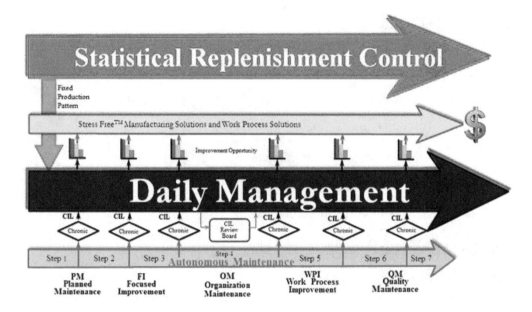

In Summary:

The daily management system and the daily leadership tour form the center or power rail of continuous improvement.

Money to the bank, manufacturing and work process solutions workshops along the improvement journey create a pay as you go improvement process.

The application of continuous skill improvement, root cause problem-solving and technical skills development all combine to increase the capability of the supply chain personnel.

A statistical inventory replenishment system using a fixed production pattern to replenish the shipped product provides production stability.

Tools for Chapter 6:

1. Standardized Daily Management

2. Check, Act (PDCA) cycle

3. TPM – Total Productive Maintenance (TPM) organized in Pillars.

    AM – Autonomous Maintenance

    PM – Planned Maintenance

    FI – Focused Improvement

    QM -- Quality Maintenance

Leadership and Daily Management Tips:

Short Term:

- Do a Stress FreeTM Leadership Capability Assessment

- Do a Stress FreeTM Organization Capability Assessment

- Implement a Stress FreeTM Daily Management Workshop.

Longer Term:

- Engage the Stress FreeTM Leadership Workshop Series

    (See chapter 7)

Chapter 6: Learning Points

- A structured, standardized daily management system is fundamental.

- The PDCA cycle is used to manage the daily management process

- Synchronized Daily Management along the entire supply chain is a fundamental element

Chapter 6: One Point Lesson - Daily Management

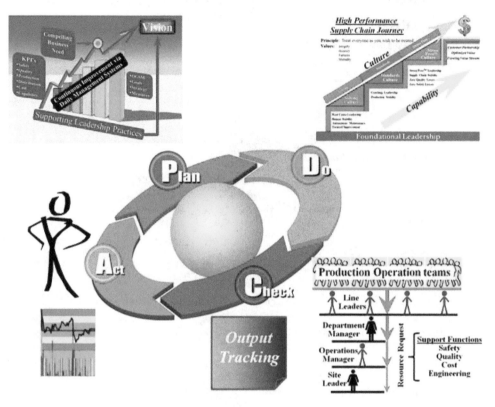

- Output Tracking begins the cycle
- Immediate response by the team to issues
- Longer term resource needs get planned
- The teams deliver to plan

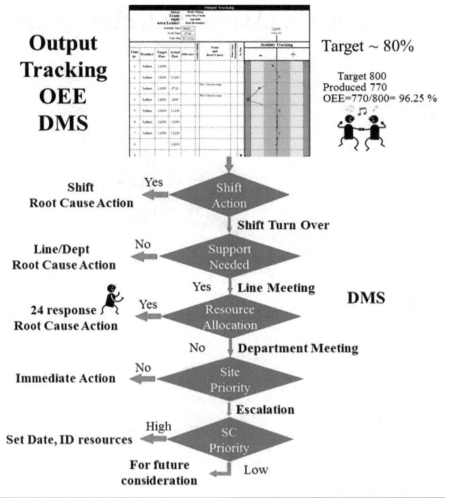

Chapter 6: One Point Lesson – Output Tracking and Daily Management

- Line output tracking begins a problem resolution resource allocation process that goes all the way up to the supply chain level if necessary.

- It stops at which ever level has the resource to solve the problem needing the resource. The Daily Management meetings are the points where the problem gets discussed and the solution resource gets allocated.

- The least that is done is to ensure the next twenty-four-hour window is address.

Chapter 7: Putting it all together

The supply chain gets established to meet the business situation existing at a specific time. It may have been perfectly designed but more than likely it was based on the best compromises available at the time. With time the supply chain was changed based on new supplier contracts, changing customer demands and the transformation process was changed to reduce cost and improve profit.

The Supply Chain leader is faced with the challenge of staying competitive, flexible, and continuously improving. The leader looks along the supply chain and lists the following tools available for use:

- Leadership hands on development
- Qualitative Assessments
- Quantitative Assessments
- Root Cause Problem Solving
- Work Process Improvement
- Rapid Change Over

Leadership hands on development

The supply chain leadership development is narrowly focused to the behaviors required to support the organization and the people being led. There are four specific leadership practices.

1. **Supply Chain Organization Assessment and Enhancement**

Every Leader needs to be out on the "floor". They must see, hear, smell, and feel the environment. They must engage the people and understand the environment. They must ensure safe, healthy work processes and a supportive environment is present.

The on the floor assessment experience and personally practiced enhancement approach, in partnership with the on the floor champions assures the success of the resulting improvement.

2. **Stabilization Behavior**

The workday actions at each level of each organization is clarified and the "on the floor time" organized and standardized. This allows the supply chain organization leaders to work together. The Organization behavior and requirements for a "successful twenty-four hour" becomes the evaluation criteria.

A shared continuous improvement application plan provides the basis for continuous improvement.

3. **Synchronization of Leaders**

Synchronization of the key supply chain work, communication and material flow is the next step toward the continuous improvement organization.

Leaders at all levels and along the supply chain make sure that everyone is operating to standards.

Reapplication along the entire supply chain is the practice.

Leaders reassess and adjust their organization to reflect increased organization capability.

4. Pull Activities and Behavior

Pull may not be achievable by the supply chain but high-quality products flowing close to the customer desired rate is possible.

The leadership behaviors supporting this environment focus on waste elimination, supply chain synchronization, organizational partnering and the empowerment and enablement of all persons in the supply chain.

Qualitative Assessments

There are many types of assessments. The ones described here are simple, facilitate a close to the action assessment, and are effective and usually only takes a half day when facilitated by an expert resource.

Comprehensive Supply Chain Assessment

Understanding the situation is critical. The supply chain crosses multiple organizations and though they may be in constant contact, leaders seldom take the time to gain a common alignment on the issues before them.

A comprehensive supply chain assessment utilizing key questions that expose the performance gaps in the current supply chain forms the basis for a common alignment. Specific countermeasure actions are identified, and a preliminary action plan is developed. This assessment is often extended to facilitate more in-depth planning for implementing immediate action.

The opportunity may be immediately known, and a specific action taken.

Additional rapid assessments may shed more light on specific opportunities. These assessments are:

Leadership Capability Assessment

Leadership behavior sets the culture. This assessment identifies the gaps in the key behaviors leaders' desire. If an organization does not have the culture leaders say they want, it most often is the result of rewarding the wrong behavior. This assessment identifies this situation.

Supply Chain Organization Assessment

The supply chain organization achieves exactly what it is designed to deliver. If it falls short, it is often due to the supply chain organization design. Leaders must design their supply chain organization to deliver the business results and the culture required to have a prosperous company.

The supply chain organization assessment identifies the gaps as compared to the winning world class design and guides the organization in developing countermeasure activities.

Quality Assessment

The customer buys the product because they inherently see the product quality value. The Quality Assessment provides insight to the gaps in the key elements that maintain the desired product quality level. The Assessment points to specific actions aimed at the elements that maintain the quality the customer desires and pays for.

Equipment Assessment

Equipment is the primary material to product transformation work agent. Keeping it performing and improving on the design is a fundamental activity of the production system.

This assessment highlights specific daily and long-term maintenance system gaps and links countermeasure actions to them.

Work System Assessment

People do work. They maintain and keep equipment and systems working and performing as desired. People grow, people contribute improvements, people represent a malleable resource that need to be supported, developed, encouraged, and empowered.

The Work system assessment looks at the specific work of the organization or work system being studied. The identified gaps can often be addressed immediately following an assessment.

Dramatic productivity results have consistently been achieved.

Safety Assessment

Safety is the number one consideration for every organization. This assessment examines safety behaviors, practices, and safety systems. The safety gaps are identified, and the leaders of the organization are guided in their safety improvement countermeasure plan.

BOSS

Quantitative Assessment

Business Opportunity Statistical Survey (BOSS) (see Chapter 3)

Inventory is the result of the current system capability. The analysis of the past year's daily product shipment tells the customer behavior story. The analysis of the past year's daily inventory position tells the supply chain behavior story. Knowing how to combine the two to determine the optimum way to synchronize product flow to customer demand in any system defines the improvement opportunity.

This analysis applied to key distribution points along the supply chain provides critical insight to the studied situation.

Two repeatedly used capabilities.

Assessments provide knowledge the root cause problem-solving and work process improvement are fundamental skills that support all other actions.

Root Cause Problem Solving

There are many root cause problem-solving tools. The goal is to have one root cause problem solving process that can be applied by those champions dealing directly with the material to product transformation and also be used by the top leader in the company.

A single process for the supply chain that meets the line to the board room criteria creates a powerful communication bond and understanding for each in between organizational layer.

Work Process Improvement

Work is synchronized to the customer need and to that work that directly produces the desired product. It is the current best and easiest way to do work. It is defined and improved by the people doing the work.

Improvement is driven by the desire to make it easier to do, the desire to eliminate the waste of time, effort, or materials. The goal becomes a mistake proof, easy to, low effort work environment.

Rapid Changeover

In today's world, the supply chain manufacturing organization must utilize their production lines to make multiple products and variations of these products. Many times, this means frequently changing their lines over to a different configuration. The time it takes to change over is time not available for production. Reducing the changeover time to that required to ensure the statistical product replenishment rate ensures the most revenue from the capital utilized.

Stress FreeTM Continuous Improvement

The following steps closely follow the content of this book. These steps define the implementation of the concepts covered in this book in a step by step process. The timing to get to step 9 is a three to five-year period. Steps 1-3 takes about twelve months. Step 4 is takes 18 to 24 months. Step 5 takes about 12 months. Step 6 -7 takes about 12 months. Step 8 is the continuous improvement state that goes on forever. Step 9 happens incrementally after Step 6, first with key customers and continues along with Step 8.

Stress FreeTM Continuous Improvement

Step 1: Comprehensive Supply Chain Assessment
Read Chapter 1

Step 2: Define the Opportunity
Improvement should be defined, planned, and tracked.
Read Chapter 1

Step 3: Envision and Empower Supply Chain Champions
Make people want to come to work. Let them experience the power of making things better. Make them champions.
Read Chapter 1 and 2

Step 4: Deploy Supply Chain Stability Actions
Create a supply chain Continuous improvement master plan as an integral part of the business plan. Operate the supply chain within a single business master plan.

Step 5: Develop the Organization
Make continuous improvement a strong pattern in the fabric of the organizations culture. It has to be more than a thread.
Read Chapter 5

Step 6: Synchronize the Work
The calendar, the clock and the control chart make synchronization possible. Technology provides the volume control and facilitates low cost solutions.
Read Chapter 4

Step 7: Create System Flow
Flow allows all supply chains to leverage their current capabilities and work toward the longer-term vision of achieving PULL.
Read Chapter 6, 7

Step 8: Create Adaptive Organization
The business world is in constant change turmoil. A flexible responsive organization will be able to adapt and ride the wave.
Read Chapter 6, 7

Step 9: Partner with Customers
The customer is the focus. An adaptive organization is capable of flawlessly dealing with the needs of the customer.
Once the customers realize that partnering is to their advantage business will continually improve.

Tools for Chapter 7:

1. The four leadership practices
2. Statistical Replenishment control
3. Root Cause Problem Solving
4. Work Process Improvement
5. Rapid Change Over
6. Stress FreeTM Continuous Improvement
7. Quality Assessment
8. Equipment Assessment
9. Work System Assessment
10. Safety Assessment

Chapter 7: Learning Points

- Daily leadership action is required
- Daily improvement reviews
- Daily Management System action.

Chapter 7: One Point Lesson - Putting it All Together

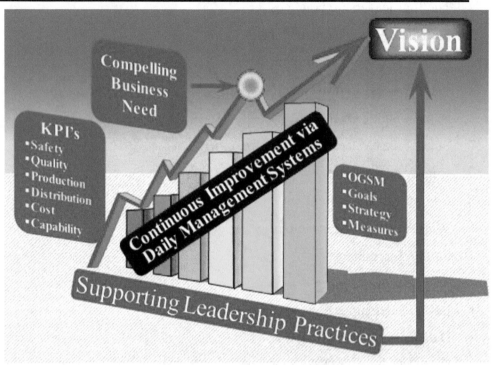

- A vision out beyond the shorter-term compelling business need

- Leadership supply chain orientation is a critical factor

- Supply Chain oriented goals with Key Performance Indicators

- A strategic, measurable action plan.

- Continuous improvement via supply chain daily management.

Ron Mueller P.E.

- Integrated Work Systems (IWS) materials author
- Coach to dozens of Manufacturing Directors across the world.
- Certified TPM Coach.
- Tested and proven to enable true breakthrough improvement of Supply Chains.

A proven leader of smart systems implementation across supply, manufacturing, and distribution to drive out cost, inefficiencies and to establish synchronized Supply Chains. He utilized the best thinking of Japan's TPM leaders and crafted the necessary related pillars and systems that work in Consumer Products Manufacturing. The results delivered include reduction of Raw and Finished Product Inventories by 40%. Delivered over $100 million is loss reduction through focused systems Workshops across dozens of sites. Developed P&G IWS program materials for external sale. Winner of P&G's Diamond Award for Contribution to Product Supply.

Core Competencies include:

- ✓ Coaching Manufacturing Leadership,
- ✓ Implementation of Integrated Work Systems,
- ✓ Statistical Replenishment design and implementation,
- ✓ Supply Chain Synchronization, author of 6 books in the Stress Free^TM series that aid Business and Supply Chain leaders to develop and improve their organization's performance.

Gordon Miller P.E.

- **Manufacturing Performance Program**
- **Development and Delivery Expert.**
- **Application of Intelligent Manufacturing technology against biggest business challenges with proven business results.**

A record as a collaborative and leading-edge thinker, developing programs to deliver cost, productivity and growth enabling manufacturing technology systems deployed via smart standards and empowered teams. As an early developer of PR/OEE measures and improvement programs, has experience with unlocking organization capability for improvement with smart strategies. Led program that developed initial P&G Manufacturing Execution System, leveraged globally across multiple GBUs. Influenced Beauty and Household Care manufacturing systems changes that enabled and leveraged global standardization for rapid footprint growth. Experience that enabled 50% reduction in OEE losses. Experience as a leader of corporate STEM talent strategy can assess and devise approaches to ensure Talent needs for the challenging future are met.

Core Competencies include:

- ✓ **Global Productivity Program Design and Management,**
- ✓ **Advanced Manufacturing Technology Innovation and Strategy Development,**
- ✓ **Development of Highly Effective Global Teams,**
- ✓ **Vendor development and management, Organization Capability Development,**
- ✓ **Talent Strategy**

Other books by Ron Mueller

Stress Free^{TM} Supply Chain Solutions

Stress Free^{TM} Manufacturing Solutions

Stress Free^{TM} Work Process Solutions

Stress Free^{TM} Changeover Solutions

Stress Free^{TM} Daily Management Solutions

This takes you to the optimum performance consulting

AROUND THE WORLD PUBLISHING, LLC,

CPSIA information can be obtained
at www.ICGtesting.com
Printed in the USA
BVHW011208020721
611051BV00010B/266